Lulu's Lemonade

by Barbara deRubertis
Illustrated by Paige Billin-Frye

The Kane Press
New York

For Julianna, who is an artist too.
—P.B.F.

Book Design/Art Direction: Roberta Pressel

Library of Congress Cataloging-in-Publication Data

DeRubertis, Barbara.
Lulu's Lemonade/by Barbara deRubertis; illustrated by Paige Billin-Frye.
p. cm. — (Math matters.)
Summary: On a hot summer day, three children squabble over which ingredients and what quantities should go into their extra special lemonade.
ISBN: 978-1-57565-093-7 (pbk. : alk. paper)
[1. Volume (Cubic content)—Fiction. 2. Liquids—Measurement—Fiction. 3. Lemonade—Fiction. 4. Cookery—Fiction.] I. Billin-Frye, Paige, ill. II. Title. III. Series.
PZ7.D4475Lu 2000
[E]—dc21
99-042677
CIP
AC

20 19 18 17 16 15 14

First published in the United States of America in 2000 by Kane Press, Inc.
Printed in Hong Kong.
GWP 1110

www.kanepress.com

"Please come down from that tree, Martin," Mattie called. "Let's think of something to do. I'm bored."

"Don't want to come down," Martin grumped. "Don't want to think. It's too hot to do anything."

Mattie's little sister, Lulu, was sprawled on the grass. "I'm thirsty," she announced.

"We could play card games," Mattie suggested.

"Boring," said Martin.

"How about watching Crazy Cat Cartoons on TV?" Mattie asked.

"Double boring," said Martin.

Mattie snapped her fingers. "I know. We could make up a recipe to enter in the Kids' Cooking Contest next week!"

"It's too HOT to cook," Martin said, swinging down from the tree. "It's even too hot to make my world famous peanut butter and dill pickle pizza."

"Thank goodness for that," said Mattie.

Lulu sat up. “You could make me something to drink. I’m really thirsty.”

“Come on, Martin,” said Mattie. “Let’s go make lemonade for Lulu.”

“Yippee!” squealed Lulu. “Lemonade!”

“Boring,” muttered Martin.

Mattie set out two pitchers. “We’ll make a whole gallon,” she said. “Which pitcher should we use, Lulu?”

“The BIG one!” Lulu answered.

“They both hold the same amount,” said Mattie. “A gallon.”

“No way,” said Lulu. The big kids were always trying to fool her.

"I can prove it," Martin said. He filled one pitcher with water.

Then he poured the water into the other pitcher. It came right up to the top.

"Wow!" said Lulu. "They do hold the same amount. I thought you were just teasing me."

Mattie was reading the directions on the lemonade mix. "This makes only a half-gallon, so the pitcher will be half full."

"Great!" said Martin. "Then we can fill it with a half-gallon of *interesting* stuff."

Lulu didn't like the sound of that. "You won't put anything yucky in my lemonade, will you?"

"Depends on what you mean by yucky," said Martin.

"You know," said Lulu.

"I don't think he does," Mattie said.

“How long will this take?” asked Lulu. “I’m really, REALLY thirsty.”

“Not too long,” Mattie said.

Lulu wasn’t happy with that answer. “Please give me some lemonade now,” she begged. “Pu-LEEEEEEEEEEZE!”

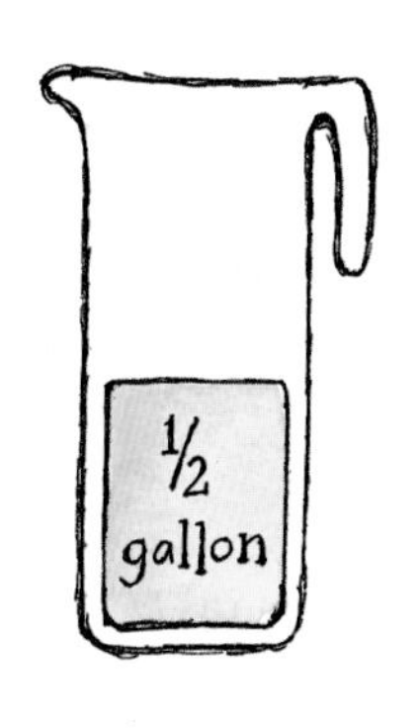

"Listen up everybody," said Martin. "We could add a quart of ketchup. THAT would make the lemonade *interesting*!"

"Yuck!" said Lulu.

"Oh, Martin!" said Mattie. "You might like that, but nobody else will."

"But everybody likes ketchup," Martin said.

"On hamburgers!" said Mattie.

The big kids were just *talking*. Lemonade was not getting made. Lulu flopped on the floor. “If I don’t get some lemonade right now, I’m going to SCREEEEAM!”

“I know it’s hard to wait, Lulu,” said Mattie. “Why don’t you draw a picture while we finish.”

Martin was looking in the fridge.

"I've got it!" he said. "Mustard!"

"Not mustard!" cried Lulu.

"Sorry, Martin," said Mattie. She pulled out a quart of orange juice. "How about this?" she asked.

"Yes!" said Lulu and Martin. They both looked surprised that they had agreed.

"Hey! Here's some lemon-lime soda pop," said Martin. "Hmmm...about 2 cups left. Just enough to add fizzy bubbles!"

"That's more like it," said Mattie. "Right, Lulu?"

Lulu nodded. "It sounds good enough to drink NOW."

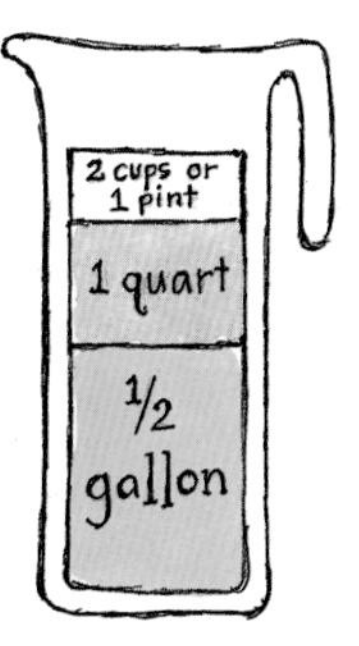

"We can't stop now," said Martin. "The pitcher's not full."

Lulu sighed.

"We can add a cup of lime juice," said Mattie. She tossed some limes to Martin.

Just then a voice called, "Can I borrow a jar of pickles?"

It was Jason from across the street.

"Sure," Mattie said. "We're making lemonade. Want to help?"

"You only like to make complicated stuff like on TV," Martin said. "Right?"

"Right!" said Jason. "In fact, Chef Fifi's show is on in a minute. We're making a pickle and beet mousse."

With a wave, Jason was out the door.

Hug
the
Cook

Lulu went back to her artwork. As she drew, she started to chant.

"Lemonade. Lemonade.
Lulu wants lemonade.
Lemonade. Lemonade. Lemonade NOW."

As usual the big kids didn't seem to hear.

Mattie was looking in the spice cupboard. "Sometimes Mom puts a tablespoon of vanilla in lemonade," she said.

"Cool!" said Martin. "I've never tried that."

"If we could just add one more thing," Martin said. He was staring at the row of spices. "A teaspoon of something unusual. Like red hot pepper. Or black peppercorns..."

“Or pepper...MINT!” cried Lulu as she ran to the herb garden.

In less than a minute, she was back in the kitchen.

"Here!" she cried.

"Just enough for a teaspoon," said Martin. He dropped half a dozen mint leaves into the lemonade.

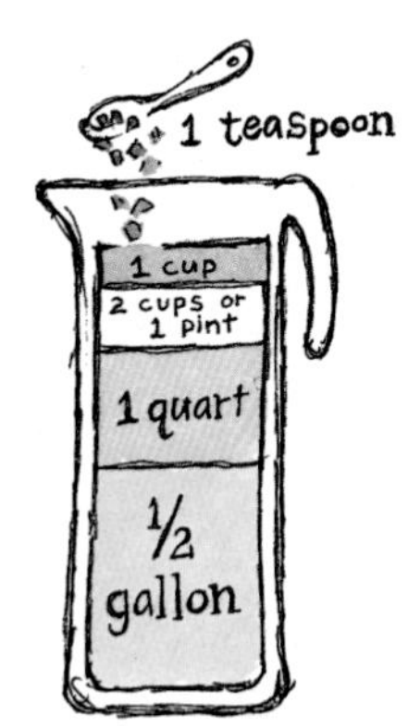

"It's ready now, right?" asked Lulu.

"Almost," said Mattie. "Want to put in the ice cubes?"

"Sure," said Lulu. She dumped the whole tray of ice into the pitcher.

"KerSPLASH! Lemonade splattered in Lulu's face.

Lulu licked her lips. "This is good!" she said. "This is the best lemonade I ever tasted!"

“Really?” said Mattie.

“Are you kidding?” Martin asked.

“Try it,” said Lulu.

They all grabbed glasses and took big swallows of the lemonade.

“It *is* the best!” said Mattie.

“It’s great!” said Martin. “I bet it’s better than anything Chef Fifi could make!”

“You should enter this lemonade in the Kids’ Cooking Contest,” Lulu announced.

Mattie and Martin stared at each other. The contest! They had forgotten all about it!

"I don't remember exactly what we put in,"said Mattie.

"Or how much!" added Martin. "You have to know that for a recipe."

"We didn't write anything down!" moaned Mattie.

"I did," said Lulu. She grinned as she held up her drawing.

Mattie's mouth dropped open. "I don't believe it," she said. "That's great!"

Martin looked stunned, too. "Good job, Lulu!" he told her.

"But what should we call our lemonade?" Mattie asked.

Lulu felt very pleased with herself. She turned over the sheet of paper. In big letters, she had written "Lulu's Lemonade!"

"A perfect name for perfect lemonade!" said Mattie.

"Except...maybe we should add just a *little* chili pepper," said Martin.

"Oh, yuck!!" said Lulu. "Not again!"

KIDS' COOKING CO

WINNING RECIPE
Lulu's Lemonade
½ gallon (64 oz.) lemonade
1 quart (32 oz.) orange juice
1 pint (16 oz.) lemon-lime soda pop
1 cup (8 oz.) lime juice
1 tablespoon vanilla
1 teaspoon mint leaves
Mix in gallon jug.
Add ice. VERY tasty.

est
lu's
nade
1
1
1
1
2
Jason's
Pickle
Torte
2

LIQUID MEASURE CHART

You can use cup, pint, quart, and gallon containers to measure liquids.

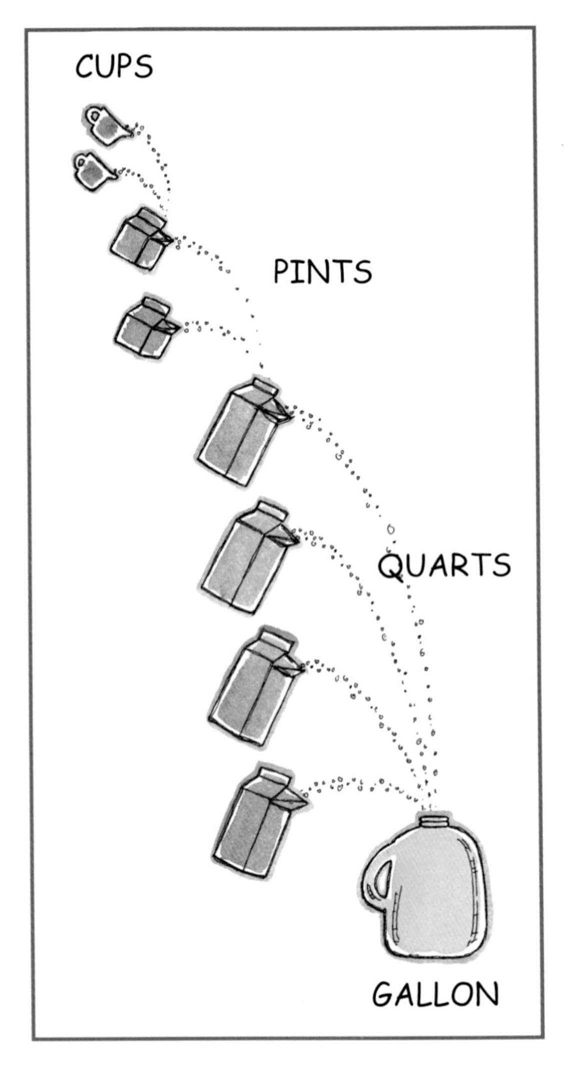

You can pour:

- 2 cups into a 1 pint container
- 2 pints into a 1 quart container
- 4 quarts into a 1 gallon container

How many cups can fill 2 pints?

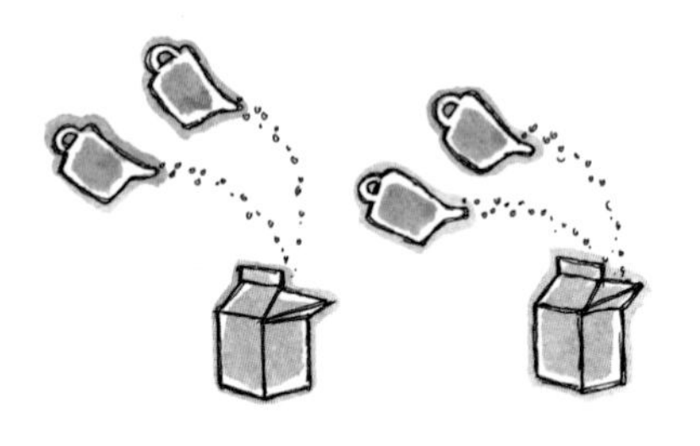

How many pints can fill 4 quarts?

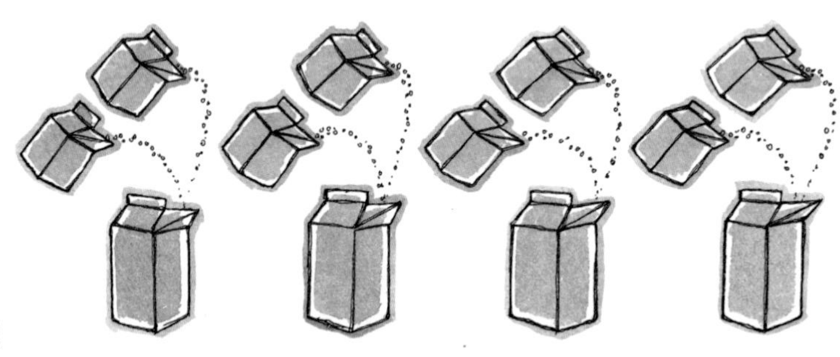